KU-084-984

Looking at Minibeasts

Worms

Sally Morgan

Belitha Press

Contents

What is a worm? 4

The worm family 6

Many segments 8

Wriggling worms 10

Living in water 12

What do worms eat? 14

Hunting worms 16

Living on others 18

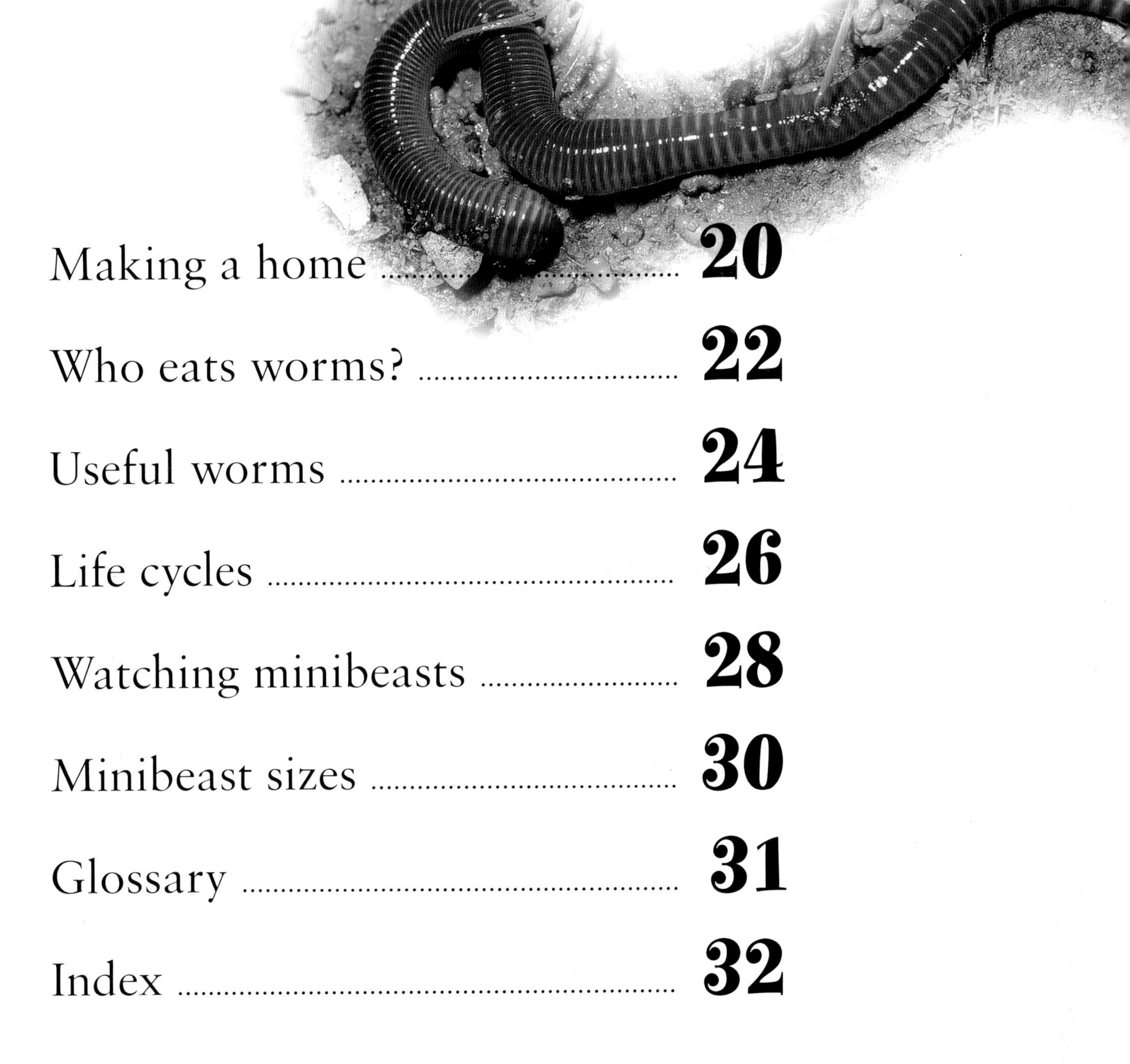

Making a home 20

Who eats worms? 22

Useful worms 24

Life cycles 26

Watching minibeasts 28

Minibeast sizes 30

Glossary 31

Index 32

Words in **bold** are explained in the glossary on page 31.

What is a worm?

A worm is an animal with a long, soft body. It does not have any bones in its body or a shell to protect it. There are tiny, **microscopic** worms that live inside other animals, and huge bootlace worms which can grow up to 40 metres long.

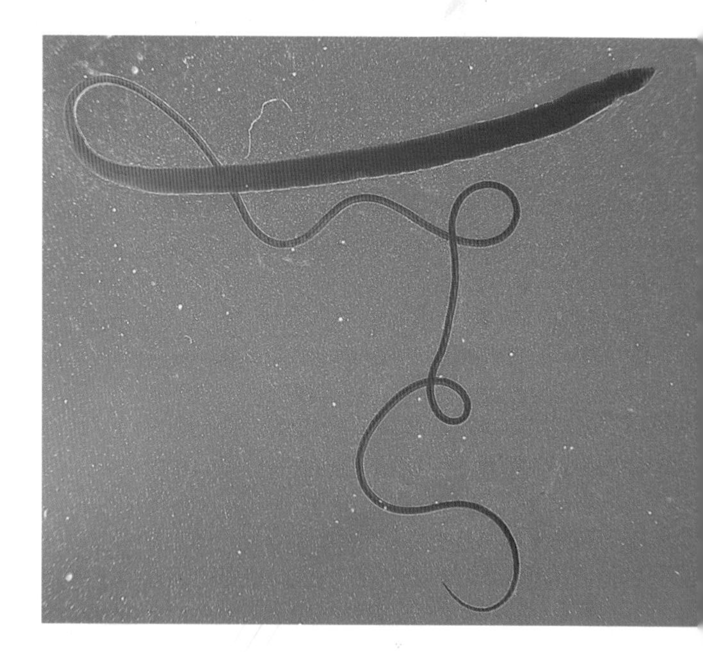

Whipworms are tiny worms that live inside the guts of other animals.

You can often find earthworms in gardens and parks.

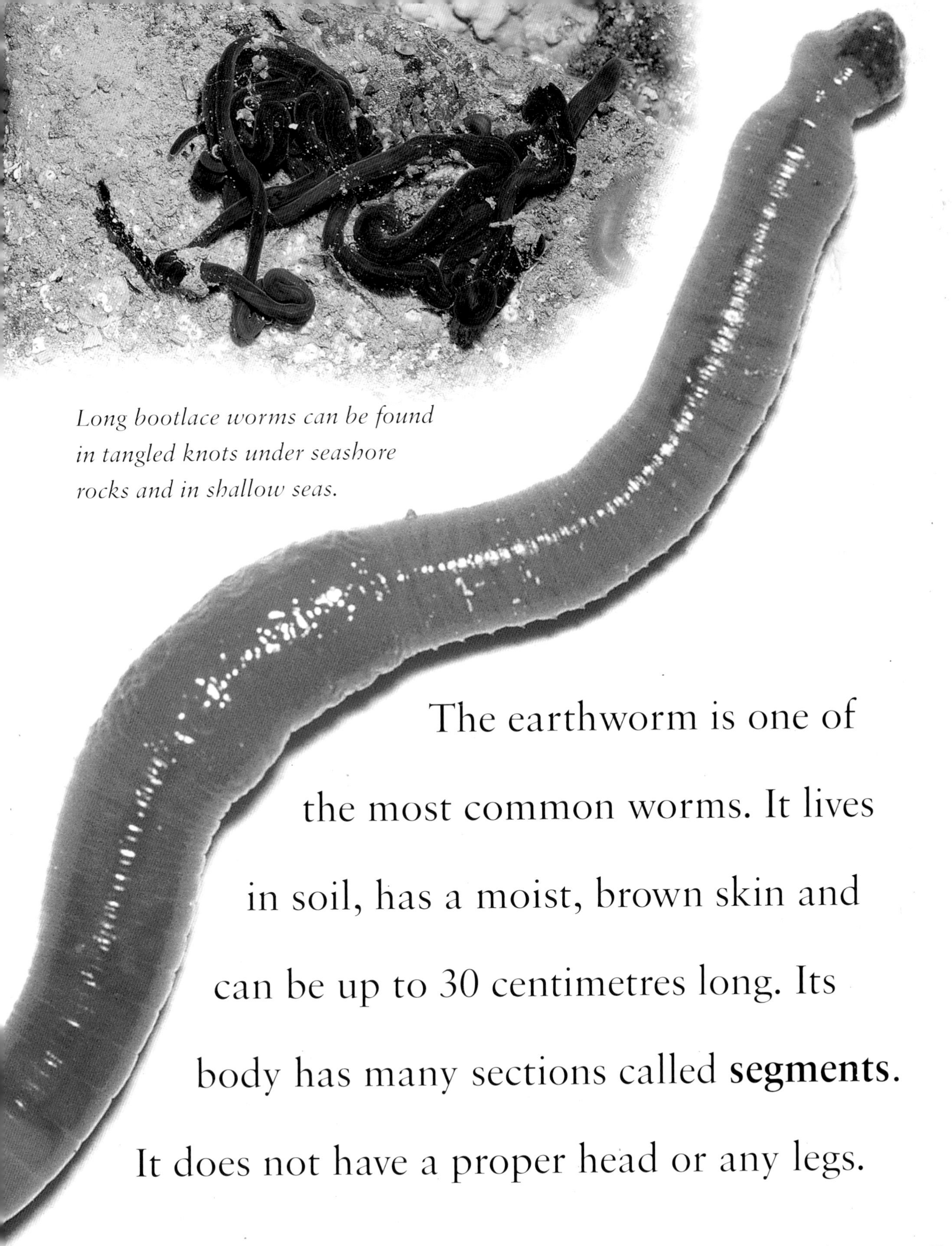

Long bootlace worms can be found in tangled knots under seashore rocks and in shallow seas.

The earthworm is one of the most common worms. It lives in soil, has a moist, brown skin and can be up to 30 centimetres long. Its body has many sections called **segments**. It does not have a proper head or any legs.

The worm family

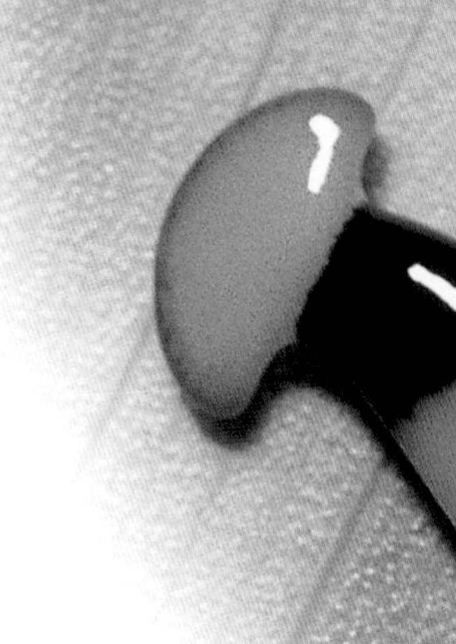

The spaghetti worm has many ***tentacles*** *around its mouth, and a long, segmented body.*

There are many different types of worms and worm-like animals. They are an amazing variety of sizes, colours and shapes.

This beautiful ribbon worm has a long body with no segments.

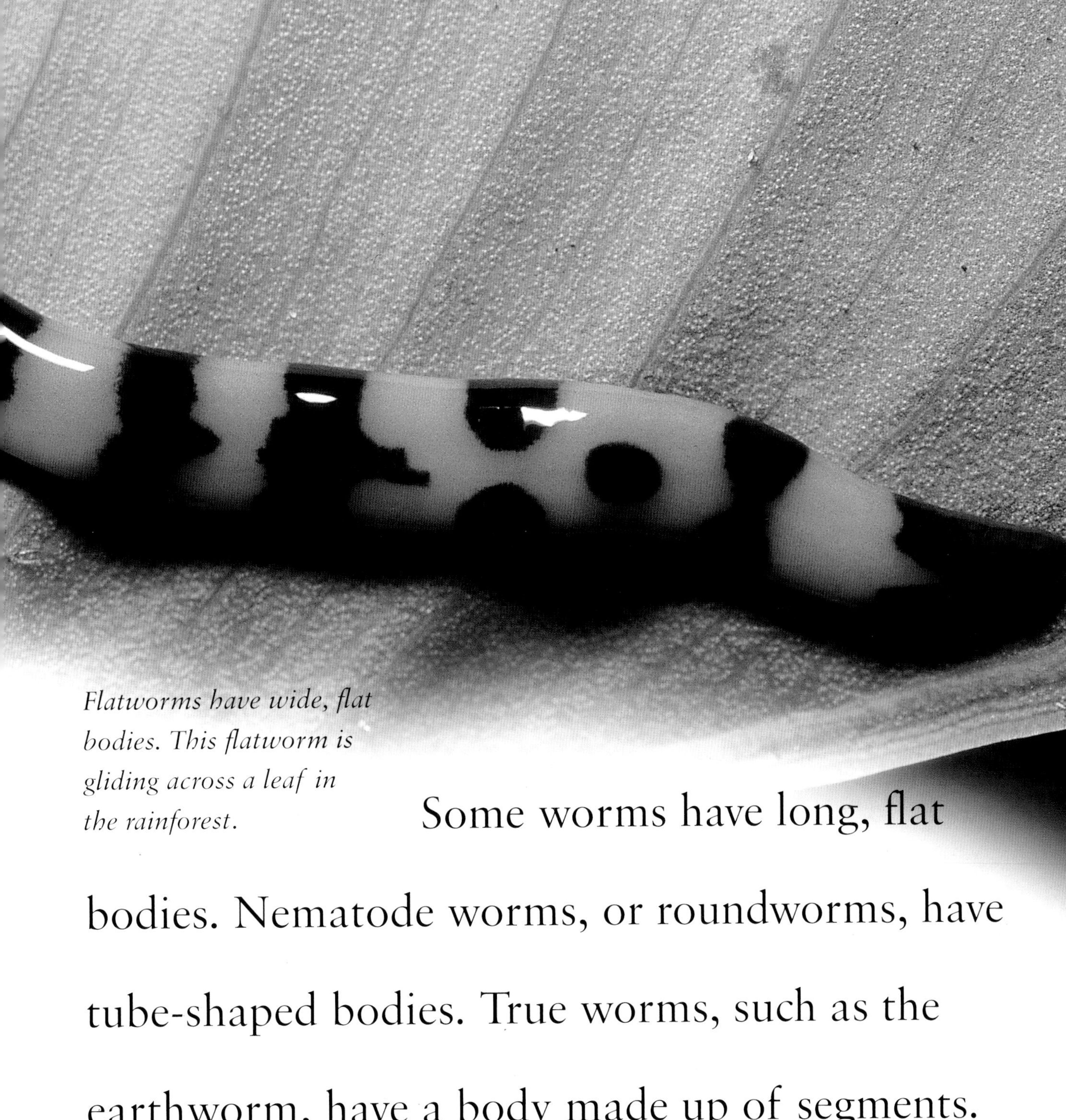

Flatworms have wide, flat bodies. This flatworm is gliding across a leaf in the rainforest.

Some worms have long, flat bodies. Nematode worms, or roundworms, have tube-shaped bodies. True worms, such as the earthworm, have a body made up of segments.

Worms are found almost everywhere, both in the soil and in water.

Many segments

A true worm, such as the earthworm, has a body made up of segments. If the end of its body is torn off, an earthworm can grow new segments. The segments have tiny hairs, called **bristles**, that help the worm to grip the ground.

The pointed end of the earthworm (left) is the head. Its mouth is on the underside of the first few segments.

The fan worm lives in water. It uses the tentacles around its mouth to catch tiny particles of food.

Fan worms, ragworms, lugworms and leeches are all true worms. A leech is an unusual worm because it has a sucker at each end of its body.

The leech uses its suckers to move along the ground and to attach itself to other animals.

Wriggling worms

The long, soft body of an earthworm is perfect for wriggling through soil. First, the worm stretches the front half of its body forwards. This makes its body long and thin.

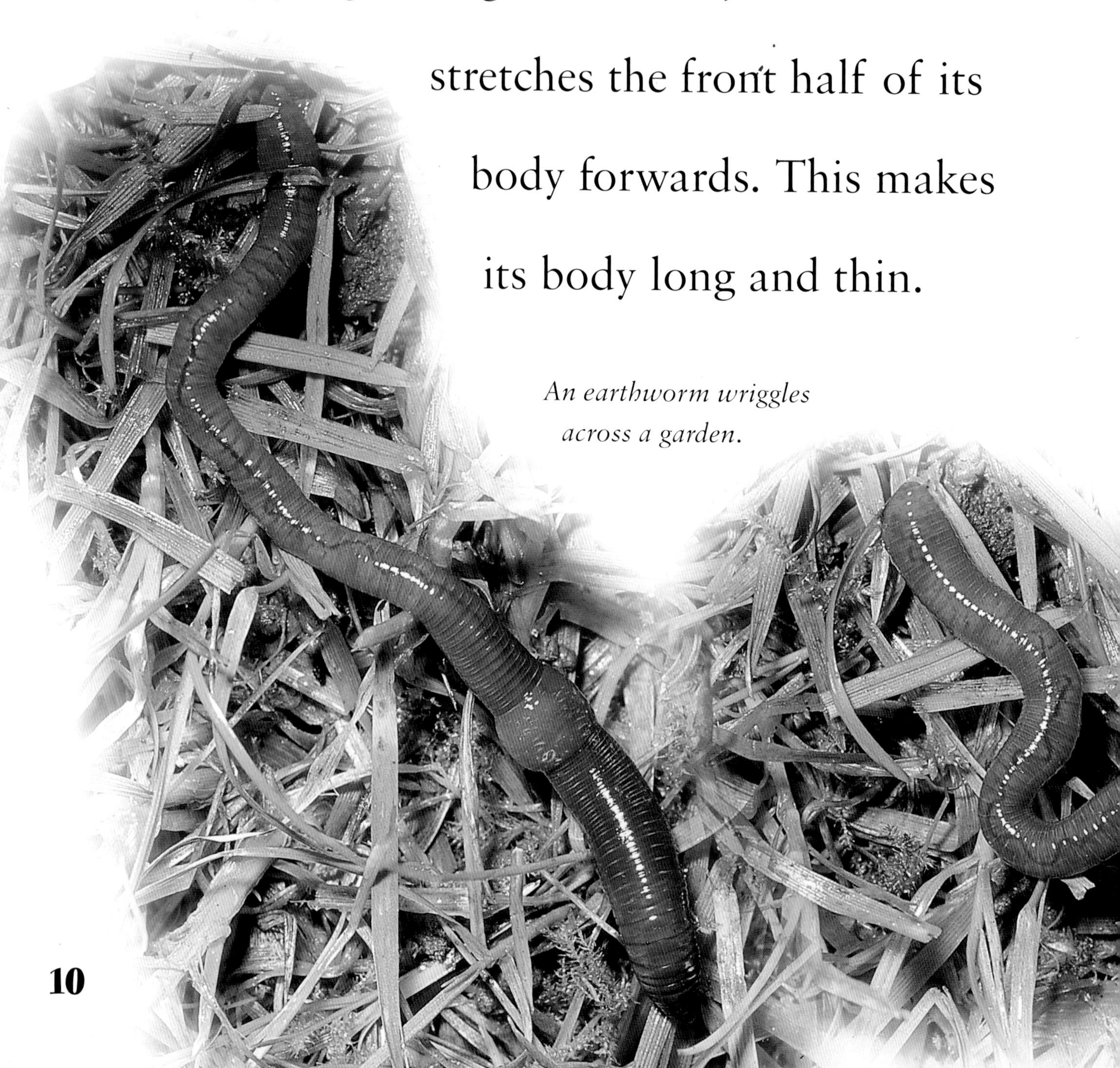

An earthworm wriggles across a garden.

Then the earthworm pulls its tail end forwards, so that its body is shorter and fatter.

Nematode worms do not wriggle in the same way as earthworms. They bend their bodies from side to side to wriggle along the ground.

A nematode worm has a long, tube-shaped body with pointed ends.

Living in water

Many worms live in water. They are found in ponds, rivers, seas and oceans.

A planarian is a type of flatworm. It glides over stones in a pond. Its large, flat body is moved along by tiny hairs on its underside.

The two spots or 'eyes' at the head end of the planarian can tell the difference between light and dark.

The paddles of a ragworm help it to crawl across damp rocks and to move through the water.

A ragworm has paddle-shaped bits along the sides of its body to push it through water. Lugworms live in burrows in the mud.

Many **marine** worms do not move. Fan worms, such as a Christmas tree worm, live in tubes attached to the sea bed.

A Christmas tree worm relies on the currents in the water to bring a fresh supply of food.

What do worms eat?

Have you ever wondered what makes the little piles of soil on top of grass? They are called wormcasts and they are the left-overs of an earthworm's meal. An earthworm eats soil and dead leaves.

An earthworm sticks its head out of its burrow to look for dead leaves on the ground.

The remains of the earthworm's food pass along its body and are pushed out to form a wormcast.

A fan worm feeds using its sticky tentacles, which are covered in **mucus**. This traps tiny particles of food floating in the water.

A fan worm feeds by stretching out long tentacles from the end of its tube.

Hunting worms

Bristle worms such as ragworms and fireworms are **carnivores**. This means that they feed on other animals. They have a small head and a large pair of jaws to catch and grip their prey.

Fireworms often crawl into clam shells in their hunt for food.

Some carnivorous worms hide in burrows and dart out to catch **prey**. They drag the prey into the burrow to eat. Others move across the sea bed looking for animals to eat.

A fireworm (below) has white bristles which inject poison.

A bristle worm (right) has long white bristles to help it move over the sea bed.

Living on others

Many worms are **parasites**. They live on or inside other animals, causing them harm.

A tapeworm lives in the gut of an animal. It attaches itself to the gut using hooks and suckers, and takes in food that passes through the gut. This means that the animal in which the tapeworm lives is starved of food.

A tapeworm is a kind of flatworm. It grows in the gut of other animals.

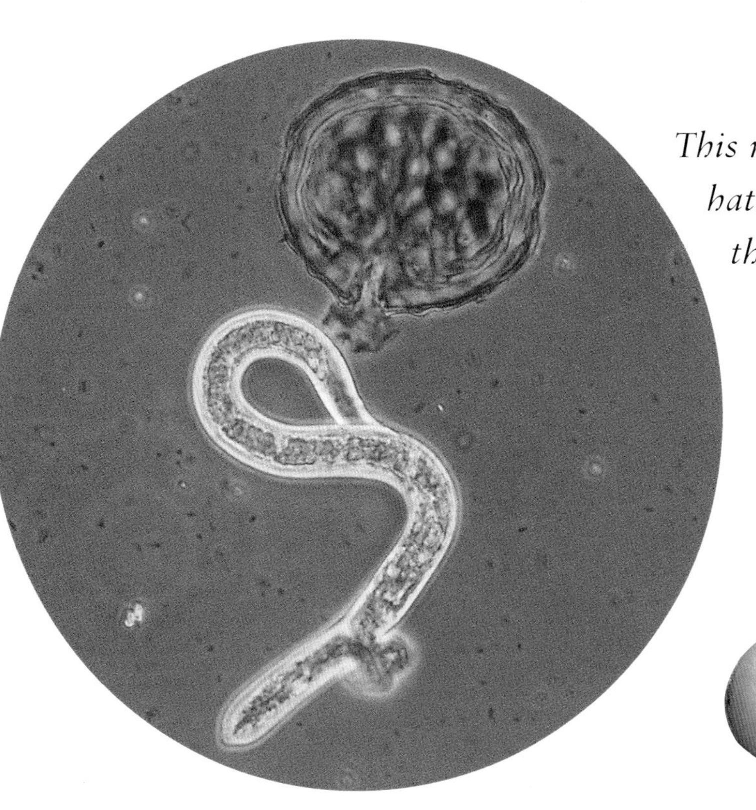

This roundworm has just hatched from its egg inside the gut of a human.

Leeches are sometimes used in hospitals. They are allowed to suck blood as part of a patient's treatment.

A leech attaches itself to the outside of animals such as fish or even humans.

It bites through their skin and sucks their blood.

When it is full of blood, it drops off.

Making a home

Earthworms live in burrows in the soil. Some earthworm burrows can be more than a metre deep. A worm burrows down by pushing its head end through the soil and swallowing the soil in its path.

This earthworm has just pushed its way out of its burrow.

At low tide, you can see wormcasts at one end of a lugworm's burrow.

A lugworm digs a U-shaped burrow in mud at an **estuary**. An estuary is a place where a river meets the sea.

A fan worm builds a hard tube to protect its soft body. The tube is attached to the sea bed.

A fan worm's tube is made from fine grains of sand stuck together with mucus.

Who eats worms?

Worms are the favourite food of many animals. But animals have to be quick if they want to catch a worm. An earthworm often keeps its tail inside its burrow. If it is scared, the worm can move quickly back into the safety of the burrow.

A pygmy shrew sniffs inside a piece of rotten wood as it hunts for worms.

An American robin pulls a worm from its burrow.

Birds try to pull earthworms out of their burrows. The worm pushes against the sides of its burrow with the **bristles** along its body. This makes it difficult to pull out. Many small animals, such as moles, shrews and hedgehogs, like to eat worms too.

Moles find plenty of worms to eat as they burrow underground.

Useful worms

Earthworms are useful in the garden. They mix up the soil as they make their burrows. They pull dead leaves into the soil too. This makes the soil richer. The burrows help to **aerate** the soil and let water drain away.

A brandling worm is an earthworm with dark and light stripes along its body. It is useful because it eats waste.

Earthworms are important **decomposers**. They eat garden waste in **compost** heaps. This helps to break down the waste more quickly.

When new waste is put on top of the heap (left), worms burrow into it to break it down. It takes several months for fresh waste (above) to be broken down into compost.

Life cycles

Earthworms mate above ground.

Earthworms usually lay

about 20 eggs at a time. The eggs are laid in a ring of sticky mucus that forms around the earthworm's body. As the worm wriggles along, the ring slips off into the soil.

The mucus hardens to protect the eggs. The eggs grow and hatch into small worms.

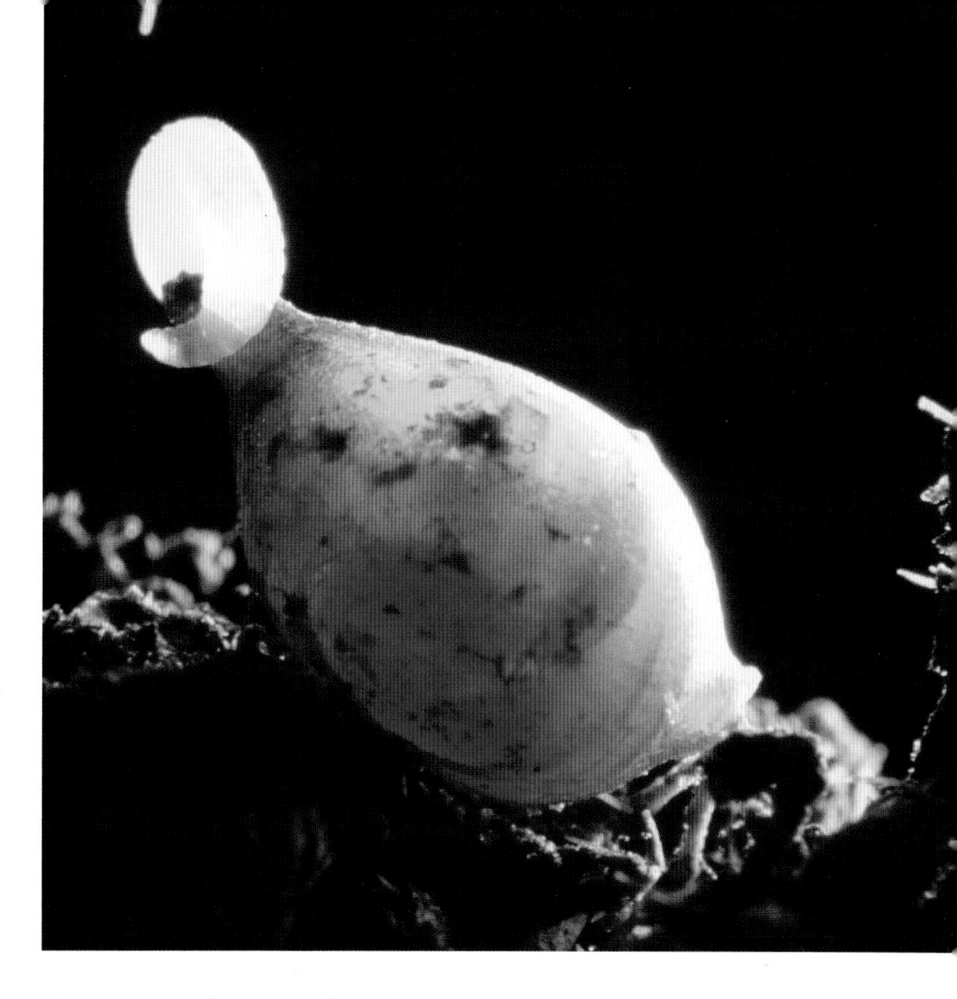

A tiny earthworm hatches from a yellow-coloured egg.

Roundworms are parasites. They lay hundreds of eggs in the gut of the animal in which they live.

Roundworm eggs are found in the soil. They are picked up by cats and dogs.

Watching minibeasts

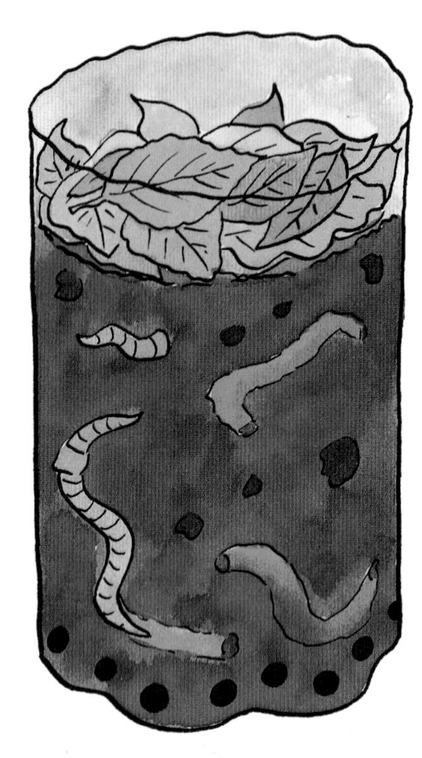

If you are lucky, some of the worms may make their burrows near the edge of the bottle.

You can keep earthworms for a short while in a home-made wormery.

Ask an adult to help you cut off the top of a large, clear plastic bottle, and to make a few tiny holes in the bottom. Pour soil into the bottle until it is three-quarters full. Then add a layer of leaves. Sprinkle a little water over the soil, and wrap a piece of black paper around the bottle.

Now place 4 or 5 earthworms on top of the leaves. Leave the worms for a few days and then take off the black paper to see what they have been doing.

Dig carefully in the compost so you don't harm the worms.

Brandling worms (see page 25) can often be found in compost heaps. Put on a pair of gardening gloves and dig down into the compost. You will probably find that most of the worms are near the top, where they feed on the fresh material that has been added to the compost heap.

In some places, there are 'worm charming' contests to see who can make the most worms come out of their burrows. See if you can make worms appear. Try jumping up and down or thumping the ground with the back of a spade.

Sometimes, sprinkling the ground with water helps to bring worms out of their burrows.

Minibeast sizes

Worms are many different sizes. The pictures in this book do not show them at their actual size. Below you can see how big some of them are in real life.

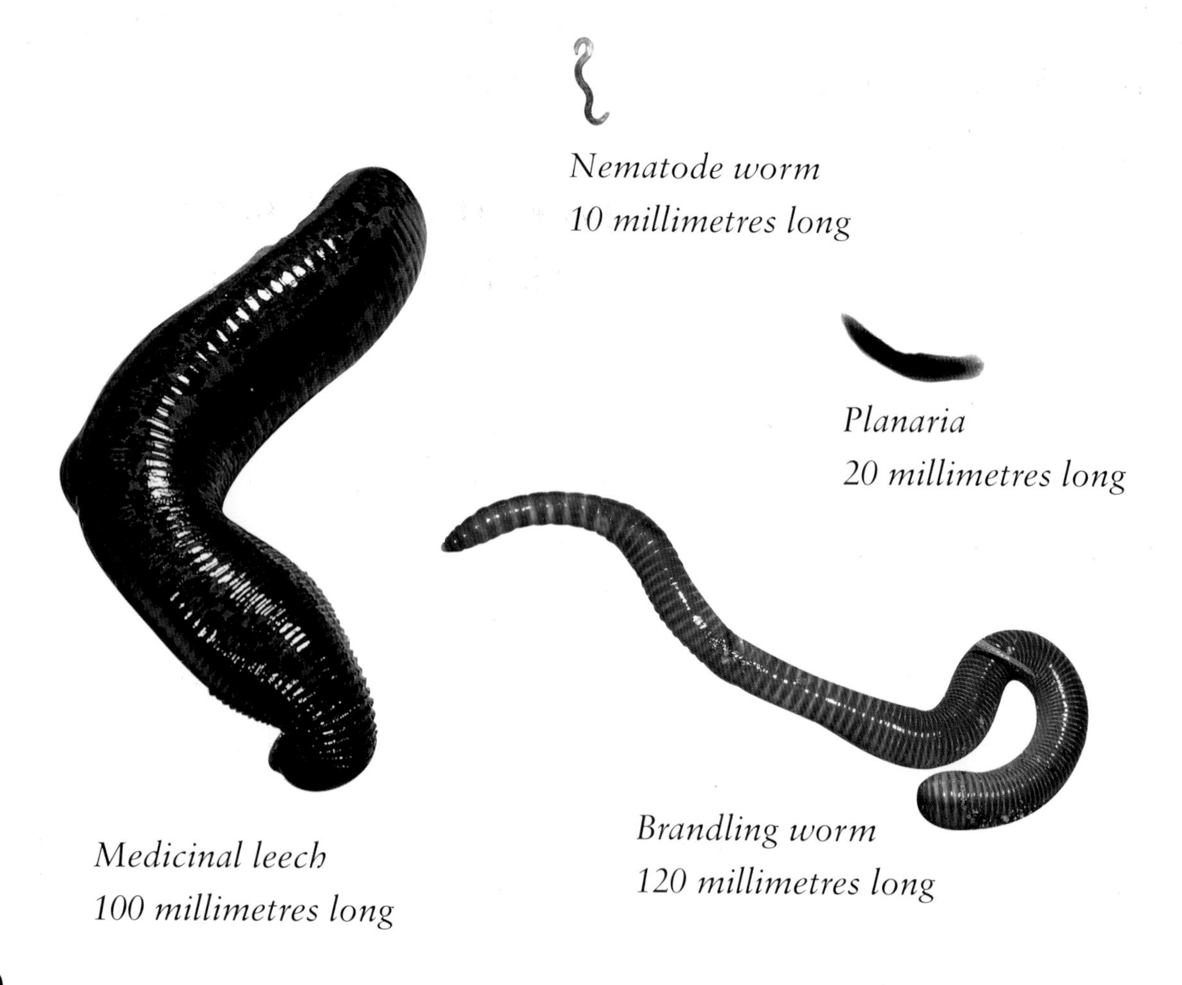

Nematode worm
10 millimetres long

Planaria
20 millimetres long

Medicinal leech
100 millimetres long

Brandling worm
120 millimetres long

Glossary

aerate To allow air to pass through.

bristles Stiff hairs.

carnivore An animal that eats other animals.

compost Broken-down plant or animal waste that can be used to make soil richer.

decomposer An animal that helps to break down dead plants, leaves and waste.

estuary The place where a river meets the sea.

marine Living in the sea.

microscopic Too small to see without a microscope.

mucus A thick, sticky substance.

parasite An animal that lives on and feeds on other living animals.

prey An animal that is killed by another animal for food.

segment A section of a body.

tentacles Long, soft parts of an animal which are often used for feeding.

Index

aerate 24, 31

blood 19
body 7, 10, 15, 23, 26
bristles 8, 16, 23, 31
burrow 20, 22, 24, 29

carnivores 16, 31
compost 25, 29, 31

decomposer 25, 31

eggs 19, 26
estuary 21, 31

food 9, 13, 15, 16, 18, 22

growing 27
gut 4, 18, 27

hairs 8, 12
head 5, 16, 20

leaves 14, 24, 28
legs 5

marine 13, 31
microscopic 4, 31

mouth 9
mucus 15, 26, 31

parasites 18, 27, 31
prey 17, 31

segments 5, 6, 7, 8, 31
soil 5, 7, 10, 14, 20, 24, 26, 28

tentacles 6, 15, 31

wormcasts 14, 21

Editors: Claire Edwards, Sue Barraclough
Designer: John Jamieson
Picture researcher: Sally Morgan
Educational consultant: Emma Harvey

First published in the UK in 2001 by
Belitha Press Limited
London House, Great Eastern Wharf,
Parkgate Road, London SW11 4NQ

Text by Sally Morgan
Illustrations by Woody

ISBN 1 84138 352 X

Printed in Hong Kong

British Library Cataloguing in Publication Data for this book is available from the British Library.

10 9 8 7 6 5 4 3 2 1

Picture acknowledgements: Steve Austin/Papilio: 23b. Scott Camazine/OSF: 24-25b. Mark Caney/Ecoscene: 16b, 17bl. Stephen Dalton/NHPA: 19br, 30bl. Robin Erwin/NHPA: 23t. Rosemary Mayer/Holt Studios: 25cl. Papilio: front & back cover tcl, front cover cl & cr, front cover b, spine, 1, 4c, 4-5, 8t, 10bl, 10-11b, 12b, 14bl, 14tr, 15b, 17r, 18b, 19t, 20b, 30cr. K. Preston-Mafham/Premaphotos: front & back cover tr, 3t, 7t, 9b, 21t, 24-25t, 26t, 30br. R. Preston-Mafham/Premaphotos: 2b, 5t, 13t. Rick Price/Corbis: 21br. Robin Redfern/Ecoscene: 22b. Kjell Sandred/Ecoscene: front & back cover tl, front & back cover tcr, 3b, 6cl, 6b, 9t, 13br. David Spears/Science Pictures Ltd/Corbis: 27tr. Sinclair Stammers/SPL: 27bl. Robin Williams/Ecoscene: 11tr, 30c.